E.H. Corson

The Star-Rider's Manual

E.H. Corson

The Star-Rider's Manual

ISBN/EAN: 9783337038021

Printed in Europe, USA, Canada, Australia, Japan

Cover: Foto ©berggeist007 / pixelio.de

More available books at **www.hansebooks.com**

THE

STAR-RIDER'S

❧ MANUAL ❧

First Edition.

An Instruction Book on the use of the American Star Bicycle.

By E. H. CORSON,
East Rochester. New Hampshire.

LANCASTER, N. H.:
THE LANCASTER JOB PRINT—A. F. ROWELL, PRINTER
1884.

G. W. PRESSEY.

(INVENTOR OF THE "STAR.")

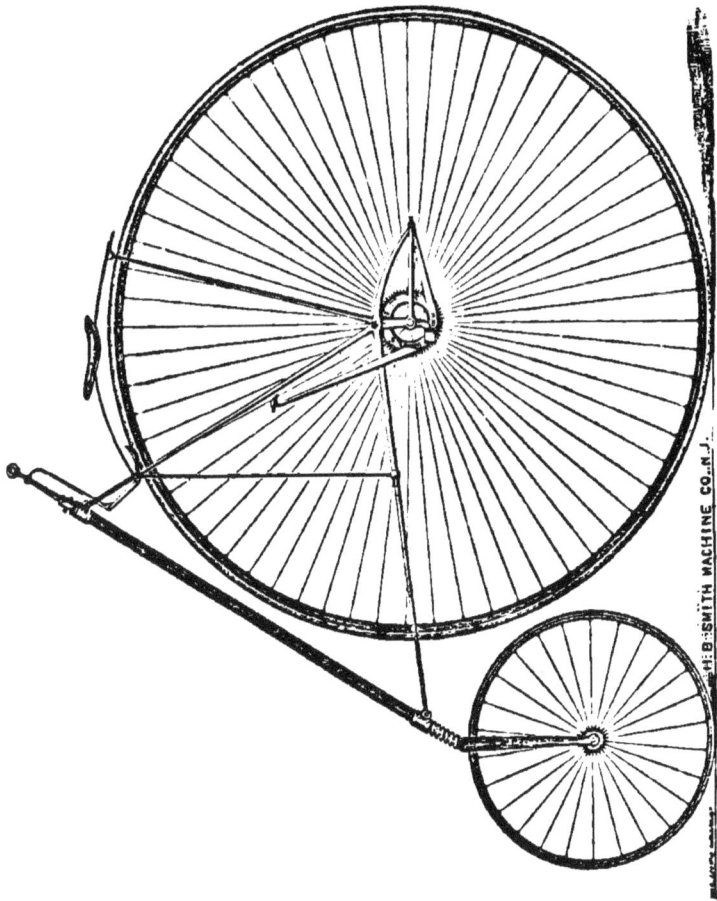

THE AMERICAN STAR BICYCLE.

(PLAIN.)

H. B. SMITH MACHINE CO., N. J.

Preface.

In offering this little work to the fraternity of Wheelmen in general, and the "Star" portion of it in particular, the author hopes to supply a want that has been growing since the advent of the new machine—a want that he felt when a beginner—and one, he confidently believes, that many others now feel, especially those who cannot obtain the advantages of a riding school. He has endeavored to lay down these instructions as concisely as possible and still preserve their clearness, thus making it a work adapted to any who may attempt to master the Star.

Besides the instructions the author has added a brief history of the Star as set forth by Mr. Pressey the inventor, and a chapter on the rights and privileges of bicyclers, from a circular sent out to members of the L. A. W., both being valuable additions to the Manual and worthy of careful perusal.

He sincerely offers his thanks to those who have assisted him in this work, and hopes to return the kindness with interest added.

Very Truly,

E. H. CORSON.

History of the "Star."

BY THE INVENTOR.

MR. E. H. CORSON,

DEAR SIR:—In answer to your request ·for me to write for publication in your "Star Manual" a history of the beginning of our new bicycle, I send you the following:

In 1879, my son Burt, then twelve years of age, was attacked with symptoms of the "bicycle fever." He borrowed a very poor specimen of a velocipede from a friend, and commenced pitching heels over head in every conceivable direction, always claiming that he "wasn't hurt much," and that his clothes were good for nothing, or they wouldn't tear so easily. He continued this exercise until oné day he took what he called a "header." This was a new term to me then and proved a very severe experiment for him, as this time he had not only torn his clothes, but had received a fearful wound in one leg from which the blood was flowing fast.

This, together with his otherwise damaged condition, convinced me that he must either have a safer machine to ride, (for I knew he would never give up riding while there was any life left in him), or that I should have but one boy, instead of two. Being a mechanic and something of an inventor, I therefore commenced studying the moddern bicycle, with a determination to make a better and safer machine for my boy to ride. In order to carry out my purpose, I myself commenced riding, and soon found that the only danger of falling, (having acquired the art of balancing, which is much the same as in walking), was from momentum, when from accident or otherwise the speed was suddenly checked; that the little wheel, which in the crank machine is only used to help the rider to keep his balance over the large wheel, was on the wrong side; that it could not serve its purpose in that position, but must be placed before the carrying wheel, where it would serve as a brace or safeguard against being thrown forward, as above. I made a machine, putting the small wheel, which I used as a steering wheel, in front, and found by careful experiments that with 150 pounds on the saddle, which could by this arrangement be placed over the center of the carrying wheel, the small wheel, with the spring above it, would run over any common obstruction, practically as easily forward as behind the large wheel, and much easier than it would as used by the crank machine, where it must carry part of the weight of the rider. Wheelbarrows are always pushed, it long ago having been found that no more power was required to push than to pull them over any common obstruction.

I then found that if I used the small wheel to steer with, the carrying wheel could be fastened firmly in its frame and save all the labor of holding it from swinging back and forth as I pressed the cranks.

Another trouble now appeared. The saddle, which was directly over the center of the large wheel, where of course it should be, was to far forward to reach the cranks conveniently. Carefully studying the crank motion, I found that a six inch crank requires about 37 1-2 inches motion to make one revolution, and when used by the foot that only about one-third of this motion was of any value, except to raise the foot to its place, and also that the power was not evenly applied; that when one crank was up and the other down there was no propelling power, while at right angles there was full power, giving a jerky and unprofitable application of force altogether too wasteful for any mechanic to use. I then tried several devices to carry the cranks past the center, but they did not fully remedy the fault. I then applied a friction clutch and lever attachment. This gave me a continuous application of power and brought the foot rests where I wanted them, and I found that with the same muscular motion required to make one revolution of the wheel with a crank, I could make nearly two revolutions with the lever, and by making the application continuous I could climb a steeper grade than I could with the crank motion. But the clutch made friction and was not durable. To remedy these faults I invented a new hub through which ran a small, stationary steel shaft. I journaled the clutch box on the hub, thus getting rid of nearly all the friction, as the box moved with the wheel when the power was applied and did not move on its journal. I then notched the surface of the propelling clutch, forming the ratchet connection which we now use, and which is very durable and gives us no trouble. By this simple and complete arrangement of parts the returning spring, driving pully and axle are all enclosed in a perfectly dust proof case. I then changed the lever, from first to second class, placing the hinge behind the axle. By this change, if the rider bears a little more than half his weight on the pedal the axle has no weight on it, thus making the machine nearer frictionless than any other bicycle in the world. And here I ought to say that in my opinion all the ball or roller bearings in use on bicycles are simply frauds and deceptions. If they are of any value, why are they not found in our machine shops and factories where the best mechanics are employed? When there is no weight on the journal, I grant they run almost without friction, but when tested under the weight of a rider down an incline, with feet off the pedals, in the hundreds of tests made all over the country, they are, beside the plain bearing of the Star, invariably left behind. To this fact some writers have answered—"your machine is heavier and of course will run down hill faster," evidently forgetting that the law of gravitation is, that falling bodies or the same density fall at the same speed without regard to size.   I next found that the bicycle was hard to control.   The staff to the little steering wheel being inclined towards the rider, the small wheel was likely to turn sideways when passing over an obstacle.   To remedy this fault, I bent the fork to a perpendicular position, when I found that the bicycle would steer automatically or, as a rider once said, "one only had to think where he wanted to go and the machine would go there."

The bicycle, being an entirely new device, had very many minor faults, which, during five years of constant study and careful experiment, we have succeeded in overcoming.   We found there were no rims either in this country or in England which would stand the tests that riders not having the fear of "headers" before their eyes, gave our machines; and at great expense, and after careful experiments, I succeeded in bringing out a new form of rim, with a flat seat for the tire and a center flange to give it stiffness, which is very much better than any other now in use. We have found, also, that a flat seated tire, while giving more bearing surface, will not roll out like a common tire of the same weight, and will run much easier.   In fact, I have endeavored to go to the bottom of the many requirements of a perfect bicycle.   The power to drive a bicycle and the resistance to be overcome in riding being fixed quantities, my study has been (1) to apply the power in a way most convenient and in the most effective manner. This I have done by placing the rider exactly over the center of one wheel.   This was an important move.   The real difference between a mordern crank bicycle and the old velocipede was that the old machine carried the rider on two wheels, the weight being about equal on both, while the new crank bicycle carries three-fourths or more of the weight on a larger wheel. This is a very great advantage, as the wheels do not always follow the same track, and if the weight is all, or nearly all, on one wheel, we have practically but one track to make.   So what was partly accomplished by that improvement, I have fully accomplished in the Star. Now with the load placed on one wheel, where it must be carried with the least power, how shall we apply the power? Not to a crank surely, for we cannot afford to make 37 1-2 inches motion to turn the wheel around once, with all the power applied on a little more than one quarter of the circle, depending upon the momentum acquired to drive the wheel over the dead centers, like a baulky horse that jumps at his load and then flies back; but the power must be applied in some way, that will give a continuous, unbroken propulsion and not require the muscular motion to be too great or too little in proportion to the power applied. To illustrate—a man who can carry 50 pounds all day, at a pace of a mile in 20 minutes and not over-exert himself, has 100 pounds to carry around a mile circle in 40 minutes. If he should carry but 25 pounds each time and make four trips of ten minutes each, his strength would be expended by too much motion and he would be much more exhausted than if he had made but two trips, carrying 50 pounds each time. So if he should carry the 100 pounds at one time and in 40 minutes, he would expend too much power and not enough motion and his strength or endurance would be exhausted very much more than if he had carried a proper load in a proper time. So I find a great waste of motion required to move the foot around the circle of the crank once for every revolution of the wheel, when, if applied to a lever and clutch attachment we can increase or diminish the motion necessary to run the bicycle a mile in a given time, by enlarging or diminishing the size of the pulley. By numerous experiments, I find that to expend the power or endurance of a rider to the best advantage, the motion necessary on a lever attachment to turn a 54 inch wheel one revolution, requires a motion of only 19 inches, or 9 1-2 down and the same back. This 19 inches motion being continuous, will enable the rider to push his bicycle up a steeper hill, showing a greater power, practically, and nearly one-half less motion than a six inch crank attachment. For proof see report of Corey Hill, and also Eagle Rock Hill up which no crank rider ever has ridden, and which a Star rider climbed without a change of machine or leverage since having won a prize for speed at Brockton. If I have preserved as much power, practically, and reduced the motion from 37 1-2 inches to 19 inches, or about 48 per cent., then I have gained in power and motion combined about 24 per cent. I do not wish to be understood to say that when we have thoroughly trained riders, and racing machines made as perfectly and as light as the crank racing machines are, we can go 24 per cent. faster, for the resistance of the atmosphere (which is the main resistance in racing) increases in proportion to the square of the speed. But I have always claimed that we have an advantage of 15 seconds in a mile.

(2) Safety. I knew the coming machine must be practically safe. It is all very well for a boy who is full of dare-devil heroism and, as the saying goes, "knows nothing and fears nothing" to declare that he cares nothing for "headers," but if we would make of the bicycle a useful carriage and a practical means of going and coming, in our intercourse with each other for business as well as pleasure, it must be safe. If it will tip over sideways, as one wheel always will, we can soon learn to make the dismount on either side with safety. The best riders, however, cannot always guard against ludicrous and even sometimes serious accidents called "headers," when the small wheel is behind. With the little wheel in front, this danger is entirely overcome and the rider has only to guard against falling sidewise. Under this head comes, too, the ability to turn short corners quickly and safely. By placing the small wheel in front we avoid the danger of falling outward, as the circle of the small wheel is outside the large carrying wheel, forming a brace just where it should be to overcome the tendency of machine and rider to fly off at a tangent. So the Star rider can turn in a three feet circle, while the crank rider, with the small wheel helplessly

dragging around on an inside circle, would hardly undertake to turn a circle less than three times that diameter. In short, we claim that the Star is a safer machine than any bicycle or tricycle now in use.

(3) A practical machine must be so constructed that anyone can ride it without respect to size. This I have accomplished by the self-adjusting swing of the pedals or levers.

(4) Speed, which is provided for as stated under the head of proper application of power and motion, and in proof of which we have only to refer to the records of Mr. Frazier, a boy only 17 years old, with one season's practice and a common road machine weighing 30 pounds more than our racing machines will when properly made, who has the best 25, 20 and 5 mile records in the United States, and the best 1 mile record, lacking one-eighth of a second.

(5) Possibilities of riding. This is a very important part of the task of an inventor who would anticipate all the requirements of a bicycle that shall claim the approval of American mechanics and inventors, and to which I have given the fullest attention. For a verification of my work in this direction, allow me to refer to my son's riding without handles, ride and running races at Springfield, fancy riding at Washington, D. C., and at many other places, for which he received first prizes; to the ride down Mt. Washington by yourself and to many other tests, which place the Star far in advance of all other bicycles for ease of control and convenience in handling.

All this careful study and experimenting, but a small part of which can be mentioned in an article like this, make up the history of the Star to the time when the perfected machines were to be put on the market.

As Mr. Edison, our American inventor, of whom we all have reason to be proud, has said, "it requires more ingenuity to get anything for an invention, no matter how valuable, than to invent it," and my own experience has not been an exception to the rule. Of the unreasonable and senseless prejudices of some of our American riders, I can only say I did not expect it, and that I am very glad to know that they are slowly but surely being overcome by facts so often proven that they cannot be denied.

That interested parties should try to belittle my invention and, by statements called "scientific," endeavor to mislead the public, I can only say I expected it, but I did not expect that they would allow themselves to make such ridiculous and unreasonable statements as they have. But writers of this kind are not always the most scientific, mechanical or trustworthy men in the world. The statement that it required the agility of a monkey and the balancing power of a Blondin to mount the Star; of another gentleman of about the same caliber, that the inventor made the handle-bar short for fear it might enter his stomach, where his dinner ought to be; another who persistently repeated a story of a "terrible header" he took on a Star, affirming that it was particularly dangerous in that respect; (Query, how could it be called a *header*, when there wasn't head enough to prevent him from telling this story?) of still another of the same class who solemnly recommends that Star riders should ride by themselves, for fear they would "tumble on to some one and injure him;" and that other *expert* in bicycle making, who states that two of the best Star riders in the country failed to keep up with the cranks and were obliged to walk up hills which crank riders climbed without trouble; and another phylosopher who pities the Star riders when they fall backward, with no hands to protect themselves with; forgetting that momentum would as surely throw the rider forward as gravitation would downward, and that if he tips backward he will always light on his feet. As all these silly statements have now been answered by practical demonstrations by the thousand riders of our machine, I only refer to them as showing one of the many unpleasant experiences with which an inventor has to contend.

To the many truly American riders and mechanics, who,

knowing the value of the scientific application of mechanical principles contained in the invention, have always been ready with a kind word for the American bicycle and its inventor, I can truly say—their discriminating and intelligent approval of my work forms the pleasantest part of my bicycling experience. With a firm resolve to push forward, improving the Star until it is perfected in all its parts, material and workmanship, and is acknowledged by all the countries of the globe to be the highest and most perfect conception of its kind, I am,

Most Respectfully,

Hammonton, N. J.,   G. W. PRESSEY.
Nov. 15, 1883.

Directions for Learning to Ride.

In learning to ride the American Star bicycle, it is important that *moderation* be used in the initial efforts, and that the learner should not become in the least fatigued, as it increases the difficulty of learning to a discouraging extent. He then becomes flurried, misapplies his efforts, and loses control of the machine. He should in all cases discontinue his practice before reaching this stage. Short practices and short rests, keeping the learner's body and faculties fresh, will advance him more rapidly than persistent efforts misapplied, and he will derive an enjoyment even in learning by pursuing this method.

The following course of instruction has been adopted, with very satisfactory results:

HANDLING THE MACHINE.—A beginner in the art of bicycle riding should watch other riders, if he can find the opportunity, and study the motions of all the parts of the bicycle and the movements of the rider until he has become somewhat familiar by observation. He should then take a machine and *handle it.* Lift it to learn its weight, and roll it about to get somewhat accustomed to the different movements and traits of his steed.

To lift the machine the learner should stand on the left side, hold it upright with the right hand on the saddle or handle, as he prefers, grasp the frame with the left hand close up to the top end of the steering shaft, then grasp the back end of the frame with the right hand, where the lever hinges; and so take it up. In this way it is carried very easily, and there is no trouble with tipping it or pinching the fingers.

To roll the machine, stand on the left side, take the handle with the left hand, place the right hand on the spring behind the saddle, or on the saddle, and push with the right hand. Roll the machine about in this manner, first in a direct line, then in circles to the right and left, guiding its course with the handle-bar. Then let go the handle and propel as before, guiding with the right hand, keeping the wheel upright for a direct line, and inclining it slightly to or from the body as it is desired to change the direction to the left or right respectively. Next walk around with one hand on either handle, first on one side then on the other. Proceed in this manner until the arms and hands are familiarized with the machine's motion.

LEARNING TO RIDE.—The learner should select as large and smooth a plat of ground as possible for the first attempts. In the absence of an instructor we would suggest that he procure the assistance of a friend who will help him hold the machine and assist him in getting on.

LEARNING TO FALL.—He should learn to fall before attempting to ride, as it will give him confidence to know that he can dismount at will or when necessary.

In getting on the machine, his assistant must stand on the right hand side and hold it upright by grasping the steering shaft with his right hand and the back end of the saddle spring with his left. The learner must take hold of both handles, (if the bicycle is too high the right hand can rest on the saddle), stand close beside the machine, place the left foot on the step, lean the machine slightly to the right and with the right leg spring up on to the step; swing the right leg around the back end of the saddle spring, drop the foot on the pedal and press it down when taking the saddle; this will start the machine. When well seated take the left foot off the step and put it on the pedal.

The assistant must let go of the machine with his left hand as the learner swings his leg around the saddle spring, and after he is mounted the assistant must change to the right side, taking hold of the back end of the saddle spring with his right hand, and grasp the stereing shaft with the left. After the learner has got the machine in motion by working his feet alternately on the treadles, his assistant must let go, first with the left hand and then with the right, and if the machine falls to the left the learner must throw his right foot around in front of the steering shaft, retaining hold of the handles with both hands, and he will alight on both feet, sustaining very little jar and saving the machine. Should the machine fall to the right, then of course the left foot should pass around in front of the steering shaft, and not over the handles. He should not jump, but wait for the machine to fall, and neither he nor the machine will sustain any injury. After he has practiced these falls or dismounts a few times so as to be able to use them when desired, then he must learn the *back dismount* by sitting well back on the saddle, taking his feet off the treadles and pulling up on the handles—the little wheel will rise and he will land on his feet, holding the machine in a vertical position before him.

Now that a knowledge of the *side falls* and *back dismount* has been acquired the learner will advance with confidence, knowing that he will not receive any harm from a fall.

LEARNING TO BALANCE—*First Ride.*—The learner must not attempt to mount alone, but let his friend assist him as before, and when well seated the assistant must take hold of the end of the saddle spring behind and assist in holding the bicycle in a perpendicular position, walking by the side of the machine while the learner is propelling it by working his feet alternately on the treadles. The more speed attained the easier it is to retain a balance. The rider must sit *erect*, and guard against swaying his body from side to side as he alternates his feet to propel the machine. He will notice a decided tendency in the machine to fall to one side. When this occurs he must *turn the small wheel in the direction it is falling* and it will immediately recover. Herein lies the chief point in the art of riding the bicycle, and the learner should thoroughly impress it on his mind. Particular care should be taken to turn the wheel *gently*, as at this stage a sudden movement, such as turning the wheel too much or lifting it from the ground, may cause an impromptu dismount. As the learner becomes more proficient, he will be able to maintain his balance by inclining the body, or pressing on the pedal opposite the side he feels the tendency to fall, or by lifting on the handles, all of which should be done with care. While thus learning to preserve his equilibrium, the learner may frequently lose his balance, and when this occurs he will come down as he has previously learned. After he has practiced the right and left curves, and can retain his position in the saddle with some confidence, he is ready for the mount.

LEARNING TO MOUNT.—This seems to the beginner to be a great obstacle to overcome, but it is not so difficult as it seems, previous preparation having taught him to balance the machine and given him confidence. It is only necessary that he should use a little care.

When learning to mount, a smooth place should be selected: one with a slight descent is best. The learner must take the machine by the handles, (if the bicycle is too high the right hand can rest on the saddle), place his left foot on the step and give the machine a little momentum by hopping with the right foot, and as he is about to spring up on the step he must lean the machine slightly from him, and when on the step, balance by turning the small wheel in the direction the machine is falling. When the machine is well balanced he should swing his right leg around the back end of the spring, put his foot on the pedal and press down as he sits on the saddle. It will be best for him to have his assistant stand on the opposite side of the machine, until the mount has been made a few times, to catch the machine in case it is thrown to the right. If he should fall to the right before properly seated his assistant will catch him, and if to the left he can easily alight on his feet. The machine is readily mounted and a few trials will suffice to bring the Star under control. If the learner has been used to riding the ordinary bicycle he must get out of the notion of leaning back and pulling up on the handle-bar when going into holes and ditches. When running up a steep incline he must throw his weight forward. In climbing hills the rider can lift on the handle-bar as much as he pleases, throwing his weight forward on the pedals. The hands should rest lightly on the handles for it takes but a slight effort to guide the Star. The tendency of a learner is to turn the small wheel too much, and too quick— a *slight* turn of the wheel will right the machine when it is falling to one side. When turning curves and circles the rider should incline his body in the direction of the curve. As a general thing a learner will lean his body away from objects he wishes to shun, and this will throw the machine towards them.

He should sit *erect* and avoid fixing his eyes on any particular object until he has obtained thorough control of his wheel. Beginners are apt to make the mistake of trying to ride on rough roads too soon—before they have learned to ride well enough to overcome the inequalities. This course should be avoided if possible, and the learner ought to be able to make the various dismounts with certainty, and *feel* that he has thorough control of the machine before trying to ride in difficult places.

Grace in riding should be studied, and the rider should sit erect on the saddle, and not move his body from side to side. The arms should be kept down and the elbows in, thus giving an appearance of ease and grace that renders the bicycle and rider picturesque and pleasing. The rider should look before him and not at the small wheel, and he should not lean on the handle-bar, but sit easily on the saddle and learn to depend on his feet, and these he should keep straight, to avoid catching his heels or toes in the frame of the machine. He should keep the *soles* of his feet on the pedals, pointing the toes down when the pedals are down, in order to get a full stroke. A full, steady stroke will give the rider more grace than a short, quick movement of the feet.

LEARNING TO DISMOUNT.—There are several ways to dismount, but for beginners the dismount by the step, and the side fall, are the easiest to acquire.

The dismount by the side fall will be used first, and has been previously described.

To dismount by the step, the rider must take his left foot from the pedal and put it down plumb under the saddle until it rests on the step, then rise up on his left foot and swing his right leg back around the saddle spring and step down beside the machine, standing in the same position as when about to mount. This is the easiest and most graceful dismount and may be acquired in a short time so it can be done when the machine is running at a high rate of speed.

Another way is to take the feet from the pedals and rest them on the frame over the ratchet boxes, then rise from the saddle and pull up on the handles, leaning back at the same time enough to raise the small wheel, then drop down behind the machine standing erect, holding the handles with the machine in a vertical position. This is a very easy way to dismount and works well when touring; when dismounting to walk up a hill or over a rough road, as the wheel will be in the best position to push, by taking hold of the back end of the saddle spring with the right hand and rolling the machine on the large wheel, raising the small wheel enough so that it will balance over the large one. Another way is to spring back from the foot rests on the frame, letting the machine run from under and catching it by the saddle spring to stop it. This dismount, with practice can be done at a high rate of speed.

A pedal dismount can be made from either side, but it is best not to use the pedals for dismounting, except when on a hard grade or in deep sand, as the lever straps might be torn from the ratchet boxes, when there is no resistance against the wheel. There are places, such as steep inclines, deep sand, bad holes and deep ditches, when the pedal dismount can be made and save a side fall; as it will often happen when riding in the dark or over rough roads, that the rider will not have time to make the step dismount.

SLOWING THE MACHINE.—This is all done with the brake, which is a very powerful one, and it can be applied with full force without danger to the rider, for he cannot take a "header" on the Star.

Wishing to "slow up," of course the rider will stop pedaling and use the brake. When riding in close order with other wheelmen, or in cities among teams and pedestrians, he should keep his fingers on the brake ready to apply it at any time.

The Star can be stopped with the brake quicker than an ordinary can by back-pedaling and use of the brake too. It takes but very little practice in the use of the brake to be able to "slow up" and stop the machine at any moment.

COASTING HILLS.—The Star leads all bicycles for coasting, and with very little practice and a proper amount of care, a beginner can coast any hill that is ridable with a carriage. July 16th, 1883, Mount Washington carriage road was coasted by the writer on a 51 inch Star.

When coasting, the rider can rest his feet on the pedals, or take them from the pedals and rest them on the foot rests; in this way he can rise up when going over an obstacle, and thus avoid much of the jar. In this way also, he can change his position as often as he wishes, which is a great convenience when making long runs or tours. If coasting very steep and dangerous hills, it is best not to let the machine run so fast that a dismount cannot be easily made in case of an emergency. By keeping the foot on the step the rider can be ready for a dismount at any moment. When coasting hills with water-bars on them, as the machine runs over the bars he must throw his weight forward. The handle-bar should be kept well in hand when coasting hills which have loose stones in the way, to prevent the small wheel turning out of its course when it strikes them. Notwithstanding the Star can be ridden down rough hills and into holes, without injury to the rider, it is very bad practice to do it at a high rate of speed, for it puts a great strain on the machine. It is better to dismount and walk down rough hills and over rough pieces of road than it is to ride and injure the machine so it will not work well on a good road. "A word to the wise is sufficient."

RIDING IN RUTS.—The Star is well adapted to rut riding, for the machine is not turned out of a straight line by propelling, as the ordinary is. The carrying wheel is held firmly in position by the frame, so the push of the rider does not throw it out of its course, if the rider sits erect.

When riding in a rut and wishing to get out, he must *lift* the small wheel out if the rut is deep and the sides are worn down square. In this way he can ride out of just as deep a rut as the large wheel will run out of. Curb stones are mounted in the same manner; also any other obstacle that the small wheel will not roll over. A rider need not have any fears about raising the small wheel, after he has had practice enough to ride well in a deep rut. When riding out of deep ruts, and up over any large obstacle, he must throw his weight forward on the pedals as soon as the small wheel is high enough, pressing most on the pedal opposite the way the machine is falling.

RIDING IN SAND.—If riding in sand or on rough roads, the rider must sit well back on the saddle, so the small wheel will be relieved of as much weight as possible; in this way he can ride through very deep sand or mud. The machine must be kept under good control and run in a straight course to accomplish the best results in sand riding.

HILL CLIMBING.—The knack of hill climbing is acquired only by constant practice. To be able to climb a long, steep hill and do it easily, the rider must have not only strong muscles and a good pair of lungs, but *perfect* control of his machine. He may be ever so strong, but if he does not have the facility of keeping his machine upright when applying full power to the pedals and handle-bar, he will not succeed in climbing a very steep hill. He must learn to manage his machine when running slow, as perfectly as when running fast, for it is best to climb a hill slowly. The rider should take it easy and use his strength to advantage. Some riders when approaching a hill will get up as much speed as possible; this will do when the hill is short, but if long and steep, it is best to approach it slowly. Rapid pedaling will soon wind a rider.

When climbing, he must keep the full power on the wheel constantly. When one lever is going down he must raise the other foot quickly and get another hold just before the lever on the opposite side is down. Herein lies one of the grand secrets of hill climbing. The rider must not get excited when approaching a hard hill; exitement will cause weakness.

The Star is a very powerful machine, and with skillful management a hill that would seem insurmountable by a bicycle can be climbed. August 25, 1883, Master Burt Pressey, a son of the inventor of the Star, climbed Corey Hill in Brookline, Mass. The length of this hill is 2,300 feet; height, 199; average, 1 foot in 11.41. Horizontal length from Prospect street, 1,464.3; height 129.3; average rise, 1 in 11.32 feet. On the last 158 feet the average is 1 in 7.85 feet, and for the next 470 feet lower down the rise is 1 in 7.87 feet. September 22, 1883, Mr. Pressey climbed "Eagle Rock" hill in Newark, New Jersey. The hill is four hundred and twenty-five feet in height, with an average grade of one foot in five. Mr. Pressey is a young lad of 17 years, and rode a 51 inch Star when climbing these hills. He is of small stature and light weight, and has ridden the Star three years. These feats will demonstrate the fact that the Star is a powerful machine, and that with practice a rider can climb very high and steep hills.

The "power trap" should be used, changing from speed to power, just as about to begin the ascent. This may be used for sand riding and in riding against the wind, to good anvantage. The trap is operated with the feet when riding. If the rider's machine is without the power trap, he must dismount and hook the lever straps back on the second pins, if he wishes to gain power. A rider must not expect to climb steep hills, ride through deep sand and in difficult places, without a great deal of practice. Practice, and this only, will make perfect.

Fancy Riding.

To be able to perform a few fancy feats on the bicycle, is an accomplishment which almost every rider would like to possess. To those who never have ridden the bicycle and to some who have ridden a great deal, it seems to be a very difficult thing to do. It is not unlike a great many other things; it cannot be done without much practice.

First, the rider must *master* his wheel; learn to ride it *well* on the road; familiarize himself with it so well that he will feel as much at home on it as though he was seated in a chair. He should not expect to do fancy riding before he can plain. Presuming that the rider has arrived at the point of perfection in riding as herein set forth, he may now go on to perfect himself in fancy riding, by learning *one thing at a time*.

The Star is capable of having a large number of feats performed with it without any falls. The rider must have his wheel in good condition, and have a short handle-bar, in order to perform all the thirty-two fancy feats which I shall give explicit instructions for, as they have been given me by Mr. Burt Pressey, the Champion Fancy Star Rider of the World.

*1. **Common Step Mount and Dismount.***—Place both hands on the handles and the left foot on the step; swing gently to the saddle without hopping. Dismount: Place the left foot on the step, and with the hands on the handles lower to the ground without jumping.

2. Step and Pedal Mount.—With the hands on the handles, the right foot on the step and the left on the pedal, swing into the saddle.

3. Still Mount.—Place the left foot on the step and the hands on the handles: lean the machine slightly to the right, then mount and start off before falling.

4. Side Still Mount.—Same as 3d feat, only place the right foot on the left pedal when mounting, and start off standing on the step.

5. Straddle Vault.—Place the hands on the handles and vault into the saddle from the ground.

6. Side-saddle Vault.—Same as 5th feat only vault in side-saddle.

7. Back Vault.—Stand behind the machine with the hands on the handles and the small wheel in the air; move forward to give the machine momentum; put on the brake and spring into the saddle as the machine tips forward; let go of the brake when the small wheel strikes the ground.

8. Back Vault—Side-saddle.—Same as 7th feat, only sit side-saddle.

9. Back Dismount.—Press on the pedals; pull up on the handles, and swing the feet to the ground.

10. Swinging Dismount and Vault.—Throw the weight of the body on the handles and swing to the ground, then vault back to the saddle.

11. Still Vault.—Same as 3d feat, only do not use the step.

12. Still Vault Sideways.—Same as 11th feat, only sit side-saddle.

13. Turning Circles.—Riding a figure 8 standing on the step with the left foot, and pedaling with the right on the left pedal. Turning small circles riding astride; sometimes turn in less than three feet.

14. Riding without Hands.—Commence on a large circle; turn smaller and smaller, then enlarge to the starting point. Steer the machine by the sway of the body and do not touch the handles.

15. Riding without Hands—Side-saddle.—Same as 14th feat, only sit side-saddle.

16. Mounting without Hands.—Place the left hand on the spring in front of the saddle and the right hand on the saddle; start the machine forward and make a regular mount.

17. Dismounting without Handles.—Place the hands as in 16th feat, and make a regular dismount.

18. Turning the Small Wheel while Riding. — Cross the arms; bear the weight on the pedals and lift the small wheel, and whirl it half around or more while up.

19. Swing from Astride to Side-saddle.—Throw the weight forward on the handles and swing from astride to side-saddle and back without touching the feet.

20. Swinging the Leg over the Handles.—Stand on the left pedal; take the right hand from the handle and swing the right leg over the bar; grasp the right handle and let go with the left hand and swing the right leg around the back of the wheel to the pedal again.

21. Still Mount from the Small Wheel.—Place the left foot on top of the small wheel and the right foot on the step; balance in this position a short time, then place the left foot on the left pedal and swing into the saddle and start off without falling.

22. Vault from the Small Wheel.—Place the left foot on top of the small wheel and balance a while; vault into the saddle and start off without falling.

23. Riding on the Steering Shaft Backwards. — When mounted and in motion, place the left foot on the step and the right foot on the left pedal; cross the arms, left over the right, then swing the left foot around to the right pedal and move ahead backwards. Return to the saddle in a reverse manner.

24. *Riding on the Steering-shaft Backwards, Propelling the Machine by the Spokes.*—Stand facing the machine, with the hands on the handles; straddle the steering-shaft and place the feet on the spokes and propel the machine by them; then place the feet on the pedals and go to the saddle as in 23d feat.

25. *Riding on the Steering-shaft, Face Forward.*—When mounted and in motion, throw the weight on the left pedal, and at the same time swing the right leg around to the left side of the machine; let go of the handle with the left hand when swinging forward of the steering-shaft, and then grasp the handle again behind the body; place the right foot on the right pedal and in this position propel the machine. Get back to place in reverse manner, while in motion.

26. *Steering with the Legs over the Handles.*—Get the machine in rapid motion, then swing the legs over the handles, at the same time letting go with the hands and steering by holding the handles in the bends of the legs.

27. *Picking up a Hankerchief.*—Drop a hankerchief, and when going past it at good speed spring backward from the machine, letting it run ahead; pick up the handkerchief and catch the machine and vault back to the saddle.

28. *Riding on one Wheel.*—Pull up the small wheel; carry the wheel well up, and when inclined to tip backwards apply the brake lightly. Steer the machine by the sway of the body.

29. *Mounting on one Wheel.*—Vault to the saddle as in 7th feat, balancing the machine as it tips forward, and riding on one wheel as in 28th feat.

30. *Mounting and Riding Backwards.*—Stand in front of the machine with the hands on the handles; place the right foot on the left pedal and make a still mount over the handle-bar to the saddle, riding backwards. While riding change to side-saddle and from side-saddle to astride.

31. *Jumping over Logs.*—Raise the small wheel over the obstruction and bear sufficiently hard on the pedals to carry the large one over.   Logs up to eight inches in diameter can easily be run over in this manner.

32.   *Jumping over the Machine.*—Take the same position as for the back vault, and spring entirely over the machine, and turn and catch it before it falls.

The preceeding list of fancy feats are performed by Mr. Pressey with as much grace and ease as a good rider would manifest in mounting and riding straight-away.

Mr. Pressey is the author, and the only one that can perform this full list of feats on the Star, at this date.   He has kindly donated these instructions in fancy riding to the *Star-Rider's Manual*, for the benefit of all who wish to follow the champion of the world in trick riding on the Star.

Touring.

Something more than a passing note should be made of this, the most delightful use to which the bicycle can be put; and not only the most delightful but the most profitable, for to tour intelligently is to gain wisdom and health, which are wealth.

The Manual on this subject, will not assume to give the best instructions for old riders to follow, but will endeavor to lay down a few rules for the novice, which if followed will give him much pleasure and some wisdom.

When a wheelman is about to start off on a tour, he should see that his wheel is in good condition and that the tool bag attached, contains an oil can, with oil, a wrench, spoke grip, cotton waste, and some small tight twisted manilla string. A bell and cyclometer should be on all well equiped bicycles, and the "Lamson" luggage carrier is an indispensable attachment when touring. The tourist's wheel being all ready, he should put on a riding suit, which must not be too heavy, for when riding he will not need much clothing. A light weight woolen shirt, knee breeches and a thin sacque coat, will make a good suit for touring. Knee breeches are best held up with suspenders. The coat should have collar and cuffs fastened to it. The stockings must be of good firm material and kept in place by elastics suspended from the waistband of the breeches.

The tourist is now ready to ride, but not prepared to stop

away from home over night. He must take along with him additional clothing, which may consist of an extra shirt, undershirt,
drawers, stockings and hankerchiefs. All may be tied up tightly
together in a roll, with comb, hair brush, teeth brush, sponge,
soap and vasaline; and around this can be rolled his coat. If
he is too warm to wear the coat he can fasten the roll to the
handle-bar by use of the Lamson luggage carrier. In dismounting at noon to sit at a hotel table, his coat may be easily assum-
ed without disturbing the inner roll. The necessities of touring
are confined absolutely to the articles which I have named, and
these can surely be carried on the handle-bar.

A good wheelman, like a good soldier, should be proud to go
in light marching order, carrying in compact form the things
that he really needs and nothing else.

With the additional clothing, when the day's ride is ended, he
can change the shirt which he has ridden in for the one in the
roll, together with the undershirt, having taken a sponge bath
and applied vasaline to any bruised or sore spots. He should
have his riding shirt properly dried during the night for use in
the next day's journey.

A low cut shoe with medium stiff sole, is, to my mind, the
best dress for the feet.

A straw hat for summer, a flat velveteen hat for early spring
and late autumn, are "Karl Kron's,"(the noted American tourist's)
preferences in respect to head covering.

The best thing the writer has seen to carry in the pocket for
drinking by the way-side is the "Tourist's Delight", (mailed free
for 25 cents) with which the tourist can drink from springs, and
the "moss covered bucket that hangs in the well," without the
danger of swallowing dirt, worms, bugs or wrigglers.

What and how to eat and drink. The item of eating and
drinking when touring is of vital importance. To know just
what to eat, and how much, in order to secure the greatest
strength and most perfect health, should be the aim of every

wheelman. Pages might be written upon this subject, but it is not the aim of the author to set himself up as an instructor in training for athletes.

Many experiments and observations have been made with the view of determining the quantity of food required by a healthy man, to maintain the most perfect health and to manifest the greatest physical power, and it has been concluded that while in active service, a man requires about thirty-five ounces of dry animal and vegetable food daily. Of this amount, ten ounces should be animal, and twenty-five ounces vegetable matter, consisting of bread, peas, beans, etc.

The training of men for pugilistic or gladiatorial contests, and for the display of great feats of physical strength, shows that our ordinary diet scale may be reduced with the greatest advantage. The amount of food required will depend, first, up-our age, second, upon temperature, and third, upon our activity. Human beings are not strong and healthy in proportion to the amount of food they eat. Our eating is one of the greatest evils in civilized life.

When starting off at an early hour in the morning, (and this is the most lovely part of the day for the wheelman to ride), the writer has found that a small quantity of oatmeal mush with sugar and cream, and some dry toast with a cup of weak coffee, or a glass of warm milk with an egg broken into it, to be a light but very nutritious meal to make a run of ten or fifteen miles on, or even more, if starting soon enough before the breakfast hour to cover a longer distance. The oatmeal mush alone will be found to make a very substantial meal. Look at the Scotch with their oatmeal porridge—as robust a set of men as ever lived. A highlander will scale mountains all day upon a diet of oatmeal, stirred with his finger, in water fresh from a gurgling spring, in a leathern cup.

After making a run of ten or fifteen miles in the early morning air, the tourist will have an appetite which cannot be acquired in any other manner.   He will eat a breakfast composed of oatmeal mush with milk and sugar, beefsteak and potatoes, corn meal griddle cakes, etc., with a good relish.   A short stop after this meal, to look around the village which the tourist may be halting in, or to view the scenery from some elevated spot, or it may be to take a sail on some beautiful body of water close at hand, will add much to the pleasures of his early morning ride.

Of course the tourist before commencing his morning run took off his cyclometer, read, and made a note of it, together with the time and place of starting, and also has made a record of the time spent in halts by the way, and remarked on the weather and road.   This is one of the many pleasures and benefits to be derived from touring; viz., to be able to tell what places have been visited, the distance between places, the time spent in riding and the condition of the roads.   The writer has found the "'Cyclist's Record Book," published by C. D. Batchelder of Lancaster, N. H., to be just the right thing for wheelmen and tourists to carry with them.   It is bound in diary form and conveniently arranged for keeping a complete daily record, and to file away for future reference.

After the tourist has made notes of what has been seen, and examined his machine to see that it is all right, he will make a note of the time and cyclometer reading, and then

> Roll along by pastures green,
> Past pleasant, ever changing scene.

The author has found the Star to be all that can be wished for as a tourist's bicycle.   It being a safety machine, the tourist can see all that he passes on either side of the road as he rides along, without being in danger of a "header".   It should not be the object of a tourist when riding for pleasure to see how far he can ride in a given time, but to ride to *see* and *enjoy* all that he can.   He will make short stops by the way when coming to particularly interesting places, and to get drinks of milk at farm houses, and a cool drink of water from the spring by the wayside, using the "Tourist's Delight" to draw the water through.

For a change the writer has tried and found the new drink, "egg-lemonade," to be not only good to quench the thirst, but to strengthen the tourist when he feels a little exhausted from a long run. Two or three kernels of roasted coffee are eaten by some long-distance riders with good results, when feeling exhausted. Sweet chocolate and chocolate candy are used by some noted wheelmen to ward off hunger when on a long, hard run, and away from places where food is to be procured.

When the dinner hour comes around, if the tourist has not been riding too much, he will feel as though he could eat his weight in almost anything, and right here he must remember one of the first cautions or hints in this chapter; viz.,"over eating is one of the greatest evils in civilized life." The well informed tourist, who is riding for pleasure and health, will guard against eating too much, and also from taking violent exercise soon after a hearty meal. He will make at least one hour's stop for dinner, and then ride easily for the first two hours after eating. For supper, he will not feel the need of eating quite so much as at dinner time, but without a doubt will have a good appetite, as a wheelman always does; so he must guard against eating too much and against eating food that will prevent his resting well during the night.

When the day's run is ended, the tourist will first clean up his wheel and put it in a proper place for the night. He will then take a sponge bath and make a change of clothing as mentioned before. This being done, his day's record should be made up in full.

Tips on Touring.

There are certain rules which every tourist should observe while on the road with his bicycle. Every intelligent wheelman will be civil, and considerate of other people's rights as well as of his own. Every person has an equal right to travel on the highways, either on foot or with his own conveyance, team or vehicle. This right is older than the constitution and statutes. The right to travel to market, to mill, to church, to public meetings, to visit relatives—that is, to pass over the public roads for the purpose of necessity or charity—is undisputed.

The supreme rule of the road is: Thou shalt use it so as to interfere as little as possible with the equal right of every other person to use it at the same time; and thou shalt be reasonably careful that no one suffer injury thereon by act or neglect of thine.

The drivers of horses have no peculiar or exclusive rights in the roads as against travelers by any other mode. As to riding on footpaths and sidewalks, it may be said that bicyclers, like travelers generally, have not only a right to a passage along the highway, notwithstanding obstructions, but if the middle of the road be impassable for their carriage, the side may be taken, and if the whole roadway, including footpaths, be impassable, they even have a right to turn out upon the abutting close by and pass over private land around the obstruction, provided they can do so without committing irreparable or very incommensurate damage. So that if in suburban streets or country roads the carriage track is in so bad a condition as to be difficult or impossible of passage by a bicycle, and the footpath can be taken without imminent risk to foot passers at the time, it is justifiable for the bicycler to take it.

When one passes by another going the same way, he is required to turn out to his left and pass by on the left of the one passed; and the latter is bound to offer a fair chance to go by, unless he be traveling as fast as the law allows; and even then, on request, if possible, he must let the other pass, for he may be going for a physician or on a public errand, or for other reason have a right to travel faster.

The rate of speed is another thing usually limited by ordinance. It is the duty of the tourist, therefore, to ascertain the laws of the roads on which he rides, and remembering that he is a carriage, to observe them accordingly. It is his duty also, to remember and to apply, according to his best and ever alert judgment, the highest rule of the road; viz, to use it so as to interfere as little as possible with the equal right of every other person to use it at the same time, and with reasonable care that no one suffer injury by his act or neglect.

Damages resulting from negligent or wrongful acts on the highway are to be recovered by and against the bicycler, of course, in like manner as by and against other travelers.

One other legal aspect may be mentioned here, and that is, that others are bound to observe the laws toward bicyclers, and bicyclers have the right to insist that they shall. The bicycler having, as we have seen, a good right to travel on the highway, any person who negligently causes him injury is liable civilly to respond in damages; and anyone who wantonly or maliciously or mischievously injures him, or his machine, is a trespasser, and liable both criminally and civilly. Numerous cases of this kind have arisen in this country and in England; and those who have assaulted or dangerously interfered with bicyclers, have been summarily punished by fine or imprisonment.

As to this point of riding without molestation, much of course depends upon the tact of the rider. Upon all hands the bicycler must take care; and he should supplement the observances of legal rights and duties with the gentler obligations of courtesy. It seldom costs much to be courteous, and sometimes it saves a large amount. Not only "a soft answer turneth away wrath," but a considerate word or act very often prevents wrath.

Every considerate bicycler will be courteous to every one. It is a courtesy of the road to turn out more than the law requires, to dismount rather than force out a loaded team; also to speak or sound a whistle or bell when approaching a street-crossing, or passing by pedestrians from behind, and sometimes when approaching a carriage to give the driver reasonable notice of approach. It is polite to use more care when meeting or passing ladies. Indeed a thousand courtesies of the road will suggest themselves to the careful wheelman, and will be obeyed.

The greatest demand on one's courtesy, however, comes from skittish or untrained horses and skittish or stupid drivers. With these drivers the tourist or rider must get along as best he can, remembering that it is easier and better to keep out of a fuss than to get out of one, and that a gentleman will be courteous even to the idiotic and the profane. Horses have rights and feelings which the good tourist will respect. There is no need of frightening or having any trouble with them, unless their drivers make it. The most of them are intelligent, and soon comprehend the new vehicle; once they have smelled it, and looked it over, they will seldom shy at it again; and it often repays a wheelman to take some pains to educate a young or spirited and a timid horse occasionally in this way. A bicycler should not ride by an unknown horse (unless in a city or attached to a heavy load) from behind without speaking, and should give him as wide a berth as convenient.

The voice is a great calmer; where a bell or whistle might startle or alarm, a word or two will quiet and re-assure. So when approaching a horse and carriage from an opposite direction, a word from the tourist will usually save all misunderstanding. The horse is a very expressive animal, and by observing him as one approaches, particularly his ears and nose and the poise of his head, one can tell at once whether it is necessary to speak, or to ride slowly to one side, or to dismount. Of course it is pure courtesy to dismount, and this every gentlemanly tourist will do, rather than frighten a horse.

There are a great many instances when courtesy will be the means of obtaining for the tourist favors from strangers, such as directions as to roads and distances, a drink of milk or water, information in regard to some interesting place or object, and many other things which will be of pleasure to the tourist, who will in turn answer the thousand and one questions that the curious are always ready to ask in regard to a bicycle. The writer has found that the old and the young of both sexes are always ready and anxious to ask questions about the "wheelman's steed," and that it is best to be courteous in answering them, even if they seem a little foolish to him. The bicycler must remember that once was the time when he was curious to know about the machine which he is now so well acquainted with. Under no circumstances should a rider pass on the wrong side of a vehicle; as, in the event of an accident, he thereby renders himself liable for damages. In turning a corner the rider should moderate his pace, and should give a signal, unless he can see a sufficient distance ahead to be assured that no vehicle is near, and that no foot passenger is crossing or about to cross. The ground in front of a horse should not be taken until the bicyclist is at least ten yards ahead. For night riding a lamp should be used to signify to other travelers the whereabouts of the bicyclist; and in frequented thoroughfares warning should be given by bell, or in some noticable manner, of his otherwise noiseless approach.

Care of the Star.

Although the "silent steed" is one that does not eat his price in oats, and having been obtained, costs nothing for keep, it cannot be said that he does not require care. To have him in good condition, it is requisite that his peculiar wants are properly attended to, and a well kept wheel is always a compliment to its rider.

The points that need the most attention are the bearings of the wheels, which should be kept well oiled and free from dust or grit, and the small wheel bearing as tight as practicable, without causing a binding; the spokes, which should be kept tight; the tire, which should be kept cemented in its place; the ratchets, which should be kept clean and oiled so the pawls will drop out easily; and the painted, polished, and nickeled parts, which should be kept clean and bright.

The bearings should be oiled before commencing a run of any length, and not allowed to get dry when on the road. It is best to put on only a few drops of oil at a time; a little and often, is a good rule. The spokes should be examined, and any that are loose tightened up at once. Care should be used in tightening the spokes not to get the wheel out of true, or get more tension on one spoke than others. They should not be too tight.

To clean or repair the Star, bring the small wheel up over the large one till the handles rest on the floor, and a perfect stand is formed, leaving both wheels free to work at with ease and comfort.

Cementing Tires.—Observe that to do it well and make it stay done, these are requisite conditions—that the tire and rim be *clean* before the cement is applied; that the parts to be cemented must be thoroughly *heated* to secure adhesion; that the tire be laid *evenly* in the rim, that is, with even tension; and that the cement be allowed to *set*, or become hard, before the wheel is used.

To Cement a Whole Tire.—Place tire on wheel with side to be cemented turned out; sear it slightly all around with a hot iron, in order that the cement may stick to the rubber; remove tire; pour melted cement into the felloe and distribute it evenly; replace tire and heat the felloe underneath; let stand twelve hours. In cementing, a spirit lamp is used, as the flame does not destroy the paint.

To Cement Part of a Tire.—Wipe off the dirt with a dry cloth and then carefully wash the surface to be cemented with benzine; add to the cement already in the felloe, if necessary, by breaking off small pieces from the cake and putting in place; then melt by passing a hot iron along the groove, or by holding a spirit lamp under the felloe, moving the flame from side to side, and being careful not to burn the tire. When the cement is melted, see that it is evenly distributed, and then place the tire in the felloe, and be careful to get it even. This done, continue heating the felloe with the lamp until the tire feels hot; scrape off the superfluous cement which has oozed out at the sides, and let the machine stand for several hours until the cement becomes thoroughly hard.

For small cuts, clean cuts thoroughly, and fill it with the cement, then tie shut with a piece of muslin until dry. Tire may be used in about two or three hours. For larger cuts, clean out thoroughly, and apply plenty of cement. Then tie shut with a

piece of muslin until dry. When perfectly dry, stitch the edges
(leaving 3-16 inches margin) with waxed thread (don't draw too
tightly) and a needle bent at the point, and smooth over the
cement. Tire may be used in about an hour. Not necessary to
remove tire for treatment. For cementing cuts in a tire use
Wahkun Cement.

Truing the Wheel.—To true the wheel the machine must be
turned up on the handle-bar and the wheel set in motion, then
by holding a piece of chalk so it will come in contact with the
side of the rim the places that are out will be easily found.
Commencing with the place which is most out of true, slacken the
spokes slightly that draw in the direction of the spring, and
tighten those that draw in the opposite direction. The spokes
must be turned but a little at a time. Work from side to
side. The spokes must not have too much tension on them.
Care must be exercised not to get the wheel out of round. A
buckled wheel had better be carried or sent to a repairer. The
wheel can be tested on its circumference by holding the chalk
so it will come in contact with the face of the tire. If it is out
of round, the spokes that draw on the parts which the chalk
does not hit must be loosened and the full places drawn down.

The Ratchet Boxes.—These must be kept clean, so the pawls
will play out and in easily as the wheel turns over. To take a box
off, unhook the lever strap and turn the lever back; take the
hook out of the strap and let the spring draw the strap back
around the box, which will be about one turn, then turn the nut
off that holds the frame to the axle and spring the frame away
enough to let the box off. Care must be taken to put the box
back on as it was; this can easily be done by putting it on with
the screw that holds the strap to the box in the same position as
it was. Care must be taken also, to have the pawls lie in the
right manner when the box is put back on. Give the spring
the same tension it had before taking the hook out, and then put
the hook back and hook it to the lever. The pawls should

be oiled but slightly, if any. The oil from the axle will sometimes work into the ratchet boxes and gum them up so the pawls
will not drop out easily; when this occurs they must be taken off
and cleaned; this will be known by the ratchets slipping and letting the lever down without any resistance, The machine
should not be used when the pawls do not work perfectly, or
when they miss catching, for it will have a tendency to wear the
ratchet teeth. The springs in the ratchet boxes should not be
tampered with by novices, and should they from any cause be
taken out, care must be had to put them back right, or they
may be broken.

To Clean and Polish Nickel.—First carefully wipe off all dust
and grit, and if there are any rust spots, rub them with a piece
of cloth or cotton waste saturated with kerosene oil, rubbing the
oil off with a clean piece of waste. Next take a long, narrow
piece of soft cotton cloth, (that has been in use and become
soft with wear will be best), and by taking one turn around the
parts and drawing the cloth back and forth in a sawing manner
a very nice polish can be given the nickel. When riding near
the sea coast, by going over the nickel with a piece of waste
well saturated with oil, it will serve as a protection against rust.
A nickeled machine must not be neglected for a day, when
around salt water, if it is desired to keep it free from rust.

All bright and polished machines may be kept clean by brushing the dirt from them and rubbing them with a soft cloth and a
few drops of oil; but if allowed to rust, rub them with crocus
cloth; if this is not sufficient use flour of emery cloth, and finish
with the crocus; this will restore the luster. It is better to avoid
the use of emery cloth, if possible, as in the hands of those in-
experienced in its use the surfaces may present a scratched appearance.

Painted machines can be kept clean by brushing the dirt from
them and rubbing them with a soft cloth and a few drops of oil.

A hint might be given here on the manner of fastening the

tire temporarily, if it should by chance come out when on the road, although this may be needless instruction to give concerning the Star with it's new square-seated tire which will not roll out with a good rider, if *not* cemented. A loose tire may be kept in place with a tight twisted string wound around the rim and tire. A good wheelman will see that all the nuts on his machine are on *firmly*, and that all loose joints are made tight so as not to rattle. Every bearing where there is any motion should be kept lubricated, not much oil at a time, but often. As to the best oil to use, there are some differences of opinion; but the requisites are, that it should have some body, i.e., but of sufficient consistency to lubricate well, and that it should not be gummy, i.e., liable to thicken and clog the bearings. Almost any good oil in use for sewing machines or light machinery may be used. A mixture of nine parts sperm to one of paraffine is good. It will be well to take the small wheel bearing out occasionally, and clean it with paraffine or benzine.

When not in use, the bicycle should have a place as all other things, and be kept in its place. This, of course, varies according to circumstances, but a dry carriage house, or any other similar building, will make a suitable place. The bicycle's stable should agree as far as possible with the following description : It must not be damp, or the bright parts will speedily suffer, and the saddle also will become mildewed if not often used. The room must not be too hot, as the heat will soften the cement that holds the tires and they will become loose, and the tires will deteriorate in quality. The room should be dry, cool and clean, with plenty of room, if possible, to clean and repair the bicycle. The bicycle ought always to be kept in an upright position, or the oil will work out at the ends of the bearings, and either damage the floor or run about the machine. This can be done in the same manner as when cleaning or repairing, or by nailing down two small strips at right-angles with the wall and close enough to it so the large wheel will rest against a piece made fast to the wall and notched out to receive the rim, with the two wheels sitting in the track formed by the pieces, with a trig nailed down in front of one of the wheels.   The writer has found this a very convenient method of stalling the Star, as it leaves it in a position to work on both sides and in front of it.

There is no doubt but that many other points will be coming up in regard to the care of the machine which the writer has not hit upon, but a good bicyclist will learn by experience what his machine needs and how to care for it.   Some riders seem to think that a bicycle, once bought, will take care of itself; they, however, soon find out their mistake.

Rights of Bicyclers.

The following extracts from a legal opinion of CHARLES E. PRATT, Esq., have just been published and sent out to members of the League of American Wheelmen, on the rights of bicyclers in streets parks, etc., and seem to prove that bicyclers have the same right to the use of highways or parkways that the owners of other vehicles have, and that they are not liable for damages if using due care.* We think they will be of interest and benefit to the readers of the Manual, so we give them a place.

"*All Persons may travel on a Street or Highway in their own common modes af conveyance ;* the use is general and open to all alike. When a street or thoroughfare has been created, and at least until it is lawfully discontinued, it is forever subservient to the right of every individual in the community to pass over the thoroughfare so created at all times.—[*Wager v. Troy, Union R. R. Co., 25 N. Y. 532; Imlay v. Branch R. R. Co., 26 Conn. 255.*]

And a street is a place in which all have a right to be, for streets are for the purposes of public travel; neither footmen nor teams, nor any class or variety of teams or carriages have any right of way therein superior to others; they each have the

---

*The opinion was obtained by the Pope Manufacturing Co., (and was published in *The Bicycling World*, Vol. II. No. 26, page 409, May 6, 1881, which see for a fuller review of rights in parks,) and it has had the endorsement of other good lawyers, and has been extensively used.

right in common and equally with the other, and in its exercise are bound to use reasonable care for their own safety, and to avoid doing injury to others who may be in the exercise of the equal right of way with them. In the use of the highways, each may use it to his own best advantage, but with a just regard to the like right of others.—[*Coombs v. Purrington, 42 Maine, 332; Barker v. Savage, 45 N. Y. 196; Commonwealth v. Temple, 14 Gray, 74.*]

Under these and other decisions enforcing well-established principles of law, it is clear that if bicycles are carriages and are used for travel, they and their riders are fully entitled to the streets; and if they are pleasure carriages and used as such, they are clearly entitled to share the common freedom of driveways or park carriage-ways, equally with any other form of carriage.

The use of manumotive and pedimotive carriages, to greater or less extent, is matter of record and description at least since 1769. The velocipede, in one form or other, is older than our State constitutions and city charters; and so, though not until within 15 or 20 years a frequent mode of conveyance comparatively, it is not a novelty, or an innovation, or an intruder among carriages in its use of highways. A velocipede is defined to be a species of carriage propelled by the rider. It may have one, two, three, four, or more wheels; it may be propelled by the feet or the hands, or both. The bicycle is a variety of velocipede considerably specialized, and consists of two wheels and a frame connecting them, with means of guiding, propulsion, and control, all constructed in the lightest and strongest manner consistent with safety of use. It supports and carries a rider like a carriage. It is directed and controlled along the roadway by the care and vigilance of the rider, like a carriage; and it enables the rider to travel, that is, to pass along over the roadway, more speedily and more easily, and more enjoyably, than he can go on foot, just as a horse or carriage enables him to do.

This reasoning, from the nature of the thing itself, seems needless, because the fact of its being a carriage, descriptively and mechanically speaking, is so obvious; and any discussion of the question would be unnecessary, were it not so often called in question.

A vehicle which has enabled its rider to cover unaided a distance of 1404 miles in six days, with which hundreds and thousands of travelers have made excursions and tours through every part of Great Britain and the Continent of Europe, and Southern Africa, and India, and Australia, and the West Indies, which is in use to the number of many thousands in the United States, and which every observant and intelligent citizen has seen in use upon nearly all our city streets and country roads from Bangor to San Francisco, and from Detroit to Tallahassee, is certainly entitled to be called a pleasure carriage without question.

When the steel and rubber bicycle was first imported into this country, it was claimed by the Collector of Customs at Boston that it was a machine, and by the importer that it was a carriage; a difference of ten per cent. ad valorem duty giving rise to the dispute. The question was referred to the department of justice, and Mr. Secretary Sherman, upon an opinion of the attorney-general, in the autumn of 1877, decided that it was a carriage, and so it has been considered in every court and every market in this country without question since.

The English Court of Queen's Bench, in the case of *Taylor v. Goodwin*, decided, all the justices concurring, that a bicycle is a carriage, and the propulsion of it by means of a person sitting on and carried by it is a driving of a carriage.—[*Law Journal Reports, part 6, June, 1879, Vol. 48, N. S.*

The highest courts in this country have not yet passed upon the question, though it has been raised in one or two of the lower courts, where it has been held to be a carriage, following the English law.—[*McFarland v. Browne. 1. Bicycling World, 27.*]

So that in every court and every judicial department where

the question has been raised for decision, it has been held to be a carriage; and it is of some weight that by the city authorities in this country, when the question has been brought fully and fairly to their attention for decision, they have always decided that it was a carriage.

The law committee of the city of Brooklyn, acting after careful deliberation, and in consultation with certainly competent legal advisers, made a report on 26 April, 1880, to the Brooklyn common council, in which they stated, "As a matter of legal right, your committee believe that bicycles are entitled to the use of streets the same as other vehicles, no more or less; subject to the same rules, liable to the same responsibility, and their violation to the laws of vehicles to be visited with the same penalties. . . . In all courts where the question has arisen, it has been without exception decided that the bicycle is a vehicle, and as such, has equal right with other vehicles to the use of the streets without discriminating restrictions, and that no authority exists by which the peculiar form of a vehicle for its motive power can be arbitrarily determined to the exclusion of some other particular class. Your committee believes this to be good law and common-sense." And the city council of Brooklyn acted accordingly, and removed all restrictions.—[*1 Bicycling World, 242.*]

So that as of an old and recognized class of vehicles by prescriptive right, and also as a vehicle recognized by judicial and quasi-judicial decisions, the bicycle and its rider have a clear right on the public streets; but even if it were a novelty, and neither it nor anything of its class from which it is in details a divergence had been used in this country prior to the summer of 1877, it would still be entitled to the use of the public streets.

Said Caton, C. J., in a leading case in the Supreme Court of Illinois, "A street is made for the passage of persons and property, and the law cannot define what exclusive means of transportation and passage shall be used. . . . To say that a new mode of passage shall be banished from the streets, no matter how much the general good may require it, simply because streets were not so used in the days of Blackstone, would hardly comport with the advancement and enlightenment of the present age.—[*Moses v. Pittsburg, etc., R. R. Co., 21 Ill. 522.*]

Mr. Justice Cooley, of Michigan, in his standard work, has found and expressed the law to be, "When land is taken or dedicated for a town street, it is unquestionably appropriated for all the ordinary purposes of a town street; not merely the purposes to which such streets were formerly applied, but those demanded by new improvements and new wants.—[*Cooley, Const. Limit. (4th Ed.), 694.*]

It is clear therefore that bicyclers have a right to the use of streets with their bicycles, subject to the same restrictions and regulations, and under the same general principles of law as are applicable to other carriages.

It is also clear that they are entitled to the same freedom of the carriage-ways in parks, unless there are restrictions under some legal authority vested in the Boards of Commissioners for their exclusion, not applicable to the case of other streets.

And this brings me to a consideration of the authority and powers of Boards of Commissioners of parks as to the point in question.

The general purpose of parks is the promotion of the health and happiness of the public; and within this general, are several distinct and special purposes of the appropriation and dedication of parts of it: parts for ornament and vegetation; parts for walking, that is, for foot-ways; parts for riding, that is, for animal ways; and parts for driving, that is, for carriage-ways.

The uses of some parts may be more restricted than those of others, according to their purposes; but all the parts must, for all the respective purposes to which they are dedicated, remain free and common to all the people.—[*Langly v. Gallipolis, 2 Ohio St. 107.*]

The *ways* in parks are of the nature of highways; at least, to the extent that is consistent with their respective uses and appropriations. The foot-paths are highways for pedestrians; the the carriage-ways are highways for pleasure carriages, at least. They are streets, and the circumstance of their being within the limits of a park or "public square" does not alter the effect.— [*Commonwealth v. Bowman, 3 Pa. St. 203.*]

Public squares and highways, streets within parks and without them, belong to the public, and are under the control and regulation of the Legislature exercising the sovereign power of the State, either by general or special law. Neither the city nor the Boards of Commissioners can act otherwise than as agents of the State and within its authority.—[*Commonwealth v. Temple, 14 Gray, 74. 4 Abbott, N. Y. Dig. Rep., and St. 555 and cases cited.*]

The right of travel in the highways and streets is one of those "privileges and immunities which are in their nature *fundamental*"; the right of a citizen of one State to travel through the highways and streets of another State, for peaceable purposes and pursuits, is one of those privileges and immunities guaranteed by the Constitution of the United States (Article 4), and which no State Legislature can take away.—[*Corfield v. Coryell, 4 Wash., C. C. 380.*]

The power of State Legislature over roads and navigable waters—that is, highways by water or by land—is substantially the same, and is one of regulation and construction, and not of obstruction or destruction; and its power is to be exercised under the restrictions of the United States Constitution and Acts of Congress regulating commerce between States. It has power to provide police regulations, to govern the conduct of persons using the highways,—for example, to regulate speed of travel, manner of passing, etc.; and to repair and alter them for the public benefit.—[*Cooley, Const. Limit. 734, 741 and cases.*]

"Every thoroughfare which is used by the public, and is, in

the language of the English books, common to all the King's subjects," says Chancellor Kent, "is a highway, whether it be a carriage-way, a horse-way, or a navigable river." The law with respect to them is substantially the same. The Crown is trustee for the public, and the use of them is inalienable so long as they remain highways.—[*3 Kent Comm. 427, 432.*]

The Constitution of New York State, Article I., § 1, declares, "No member of this State shall be . . . deprived of any of the rights or privileges secured to any citizen thereof, unless by the law of the land, or the judgment of his peers." This was so in 1846, and to the present time. By the same Article it is declared that no citizen shall be deprived of life, liberty, or property, without due process of law and that no law shall be passed abridging the right of the people peaceably to assemble, etc.

Now, the privilege and immunity of the use of the highway and the public street is necessary to the enjoyment of liberty, and to the right to assemble; since there can be no liberty without action, or movement to and fro, nor any assembling without travel; and the use of the highways, being a necessary incident, is secured by the Constitution.. And by the judical decisions, the use of carriage-ways and streets includes the right to use one's own common mode of conveyance—such carriage as he has.

I have grave doubt whether any Act of the Legislature of any State, prohibiting the use of bicycles under any reasonable regulations in the streets and highways of that State, or any considerable number of them located together, would be a valid or constitutional statute. In this connection the language of Gibson, C. J., in the case of *Com. v. Bowman, 3 Pa. St. 203,* is specially pertinent:—

"County commissioners have no greater right than an individual has, to disturb the citizen in the enjoyment of a municipal franchise, at least beyond the bounds of absolute necessity. . . . The public square is as much a highway as if it were a street, and neither the county nor the public can block it up to the prejudice of the public, or an individual; nor can either assert a right to it by enclosing it beyond a reasonable curtilage. It is dedicated to the use of all the citizens as a highway, and all have a right to pass over it without unreasonable let or hindrance."

One other point may be referred to, in passing, which has been allowed too much force in some quarters heretofore; and that is the alleged dangerous and unwelcome aspect of bicycling to the horse-using portion of the public. Commissioners undoubtedly have a right, under their authority to regulate, to prevent pedestrians from carrying strange illuminations, or nitroglycerine cans, or bombs on the foot-ways; because they are not necessary to the use of the foot-ways, and are unreasonably dangerous. But they have no right to prohibit the carrying of umbrellas, canes, crutches, cork-legs, or personal deformities on the foot-ways, although they were objectionable to some, and umbrellas frighten timid or untrained horses; because they are necessary to the full and free use of the foot-ways.

So they might prohibit the covering a carriage with fantastic shapes and colors, and hanging its wheels all over with bells; because that is something more than simply a carriage, and is not necessary to the full and free use of the drive-ways, and is unreasonably dangerous.

Now the bicycle is no such contrivance. It has by the prejudiced, the thoughtless, or the ignorant, been sometimes called dangerous. But it is simply a carriage (reduced to the lowest possible terms), and nothing else; it has nothing unnecessary to travel, no dangerous additions to the vehicle proper. Ridden at a rate not greater than six miles an hour, it is capable of complete control,—deviation from its course, immediate stopping, etc.; and is not as dangerous as other vehicles, for there can be no runaway or disastrous collision. It is said that horses are frightened by it. The fact is that they are not, any more than they are at umbrellas. But if they sometimes are, it is to be remembered that the highest courts have decided that drivers of horses have no more rights in streets or carriage-ways than those using other common modes of conveyance, and that the mere frightening of horses is neither actionable as a tort, nor complainable as a nuisance, nor an obstruction which city officers or public boards are accountable for.—[*Moses v. Pittsburg, etc., 21 Ill. 522; Cook v. Charleston, 98 Mass. 80; Stone v. Hubbardston, 100 Mass. 50; Keith v. Easton, 2 Allen, 552; McFarland v. Brown, 1 Bicycling World 27, and Macomber v. Nichols, 34 Mich. 212*, is a very good case in point.]

"An ordinarily gentle and well-broken horse" is the kind of horse to be considered, according to the language of the courts, and these are not frightened by bicycles. The common experience of cities and towns may be well indicated by quoting the language of the London *Standard* (not especially favorable to bicycles), in commenting editorially upon a case before magistrates relating to tolls on turnpikes in August, 1879: "'The prejudice against bicycles has all but disappeared in London; the horses are now accustomed to the machines, and are no more frightened by them than by other vehicles; the riders themselves are very careful, and the number of accidents caused by them is surprisingly small; and the people in general look with pleasure uson the flying wheels as they scud noiselessly along."

To sum up, however, upon the law as I find it, and the logic of the decisions as closely as I can apply it, my opinion is, in brief, that the drive-ways of parks are public streets for the purposes of pleasure travel at least; that all persons have a right to use the public streets with their own common carriages; and therefore that all persons have an equal right to use the drive-ways of parks with their own common pleasure carriages; that bicyclers are within that class; that the bicycle is a common pleasure carriage; and that therefore the bicyclers have an equal right to use the drive-ways of parks with their bicycles.

And further, that town or city or county governments or officers have no authority to exclude, or to regulate so far as practically to exclude bicyclers from the public streets; that the Commissioners of parks have no greater authority than such other governments or officers in the matter; and therefore, that the Commissioners of parks have no authority to exclude, or to make such restriction as to practically exclude bicyclers with their bicycles from the carriage-ways of parks."

CHARLES E. PRATT.